Gateways
to Science

Gateways to Science

Neal J. Holmes
Department of Chemistry
Coordinator of Science Education
Central Missouri State University

John B. Leake
Department of Science Education
University of Missouri

Mary W. Shaw
Department of Education
Montgomery County, Maryland

Webster Division, McGraw-Hill Book Company
New York/St. Louis/Dallas/San Francisco/Auckland/Bogotá/Guatemala
Hamburg/Johannesburg/Lisbon/London/Madrid/Mexico/Montreal/New Delhi
Panama/Paris/San Juan/São Paulo/Singapore/Sydney/Tokyo/Toronto

ISBN 0-07-029821-1

1 2 3 4 5 6 7 8 9 10 DODO 91 90 89 88 87 86 85 84 83 82

Consultants and Reviewers

Editor in Chief: John F. Mongillo
Sponsoring Editor: M. Jane Kita
Editing and Styling: Caroline Levine
Photo Editor: Suzanne V. Skloot
Production: Angela Kardovich
Editorial Assistant: Joann McPartland

Editor: Linda Nicholson
Photo Researcher: Randy Matusow
Text and Cover Design: Group Four, Inc.
Design Supervision: Frank Medina
Cover Photo: Tom McHugh/Photo Researchers
Layout and Art Production: Cover to Cover, Inc.
Artists: Tom Noonan, Sandy Rabinowitz, John A. Lind Corp.
Contributing Writer: Sarah F. Corbin
Editorial Assistant: Lorraine Rogers

This book was set in 18 point Aster by Black Dot. The color separation was done by Black Dot.

Contents

Life science

1 **Dandelions 3**

Activity—Draw a plant 8

Show what you know 9

2 **Stems and roots 11**

Activity—Make a chart 16

Show what you know 17

3 **Seeds 19**

Activity—Plant a seed 23

Show what you know 25

4   Leaves 27

Activity—Make a half leaf whole 30

Show what you know 31

5   Buds 33

Activity—Make buds open 38

Show what you know 39

6   Goldfish 41

Activity—Make an aquarium 46

Show what you know 47

7   Animal life 49

Show what you know 55

8   Ocean life 57

Activity—Make an ocean picture 62

Show what you know 63

9   **Tracks 65**

Activity—Make some tracks 69
Show what you know 71

Health and safety

10   **Senses 73**

Activity—Play a game 78
Show what you know 79

11   **Feeling good 81**

Show what you know 87

12   **Safety for you 89**

Show what you know 95

Earth and space science

13 — **Temperature 99**

Show what you know 105

14 — **Water in air 107**

Activity—Where does water go? 110

Show what you know 113

15 — **Day and night 115**

Activity—Make a chart 119

Show what you know 121

16 — **Seasons 123**

Show what you know 129

Physical science

17 — **Mirrors 133**

Activity—Working with mirrors 138
Show what you know 139

18 — **Shadows 141**

Activity—Does a shadow move? 144
Show what you know 147

19 — **Who is right? 149**

Activity—Is it big or little? 153
Show what you know 155

20 — **Magnets 157**

Activity—What can magnets do? 159
Show what you know 163

21

Activity—Take a closer look 168

Show what you know 171

Discover more 172
Vocabulary list 174

1

Life science

1. Dandelions
2. Stems and roots
3. Seeds
4. Leaves
5. Buds
6. Goldfish
7. Animal life
8. Ocean life
9. Tracks

Health and safety

10. Senses
11. Feeling good
12. Safety for you

1

1

Dandelions change.

What changes do you see?

Find some dandelion **seeds**.

Where are the seeds?

What color is a seed?

Blow on some dandelion seeds.

Where will they go?

Activity

Draw a plant

Draw a dandelion.

Color it.

Name the flower, leaf, and seeds.

Write the names on the drawing.

Show what you know

Look at the flowers.

Find the dandelions.

2

Stems and roots

Dig up a plant.

Part grows under the ground.

It is called the **root**.

Part grows above the ground.

It is called the **stem**.

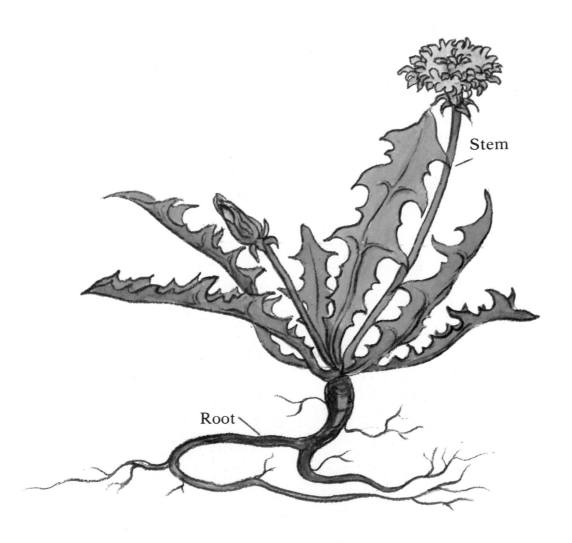

Stem

Root

Most plants have stems and roots.

Find some stems.

Find some roots.

People eat the stems of some plants.

People eat the roots of some plants.

Animals eat stems and roots, too.

Vegetables are plants we eat.

Can you name these vegetables?

 **Activity**

Make a chart

Make a vegetable chart.

Look for pictures of stems and roots.

Cut them out.

Paste the pictures on the chart.

Which ones do you like?

Stems We Eat	Roots We Eat
celery	carrots
asparagus	onions

Show what you know

Look at these vegetables.

Some are roots.

Some are stems.

Some are leaves.

3

Seeds

This is an **oak** tree.

An oak tree has **seeds.**

The seeds are **acorns**.

Many acorns fall from an oak.

What happens to these seeds?

Will all acorns grow into oaks?

Some seeds grow where they fall.

Other seeds stick to animals.

Birds carry seeds away, too.

Wind carries seeds to other places.

Water carries away seeds, as well.

Many plants grow from seeds.

Look **inside** a seed.
What do you see?

Show what you know

Make a plant and seed chart.

25

4

Do both sides of a leaf match?

Activity

Make a half leaf whole

Cut a leaf in half.

Put a mirror along the cut side.

Look in the mirror.

What do you see?

Show what you know

Match each leaf with its tree.

5

Buds

Buds tell us a plant is living.

Leaves grow from buds.

Flowers grow from buds, too.

This is a bud on a **branch**.

This is the bud 1 week **later**.

This is the bud 2 weeks later.

This is the bud 3 weeks later.

Was this a flower bud?
Was it a leaf bud?

Look at the **pussy willow** buds.

These buds are flower buds.

Pussy willows are **gray** flowers.

Look at the **lilac** buds.

What will grow from them?

What makes buds open?

Buds need water and warm air.

Then they will open.

Most buds open in spring.

Not all buds grow in spring.

Some plants grow flower buds in fall.

House plants grow leaf buds all year.

Activity

Make buds open

Most buds open in spring.

Make some buds open **earlier**.

Bring them inside.

Put them in water.

Watch them open.

Show what you know

Which sentences are right?

1. Buds grow into plant parts.

2. Buds are always on branches.

3. A bud can become a leaf or a flower.

4. Buds grow only in spring.

6

Goldfish

Watch a **goldfish**.

What color is it?

41

Goldfish are like most fish.

They live in water.

They use **fins** to **swim**.

Find the fins on this fish.

Many fish have a long tail.

Most fish have **scales**.

Look at the goldfish.

Can you see its scales?

Does it have a long tail?

How does a goldfish use its tail?

Look at this fish.

It has a good shape for swimming.

Why is the bottom side light?

It is easy to care for a goldfish.

Activity

Make an aquarium

Put water in a jar.

Put a water plant in the jar.

Add a goldfish.

Feed your fish every day.

Show what you know

Which sentences are right?

1. Goldfish live in water.

2. Goldfish use their fins to eat.

3. Only goldfish have scales.

4. The bottom side of a goldfish is light.

7

Animal life

Most animals move around.

Many water animals use fins.

Most land animals use legs.

Most birds **travel** in the air.

They use their wings to fly.

How do **snakes** move?

Animals **escape danger** in many ways.

Geese fly away.

How does a turtle escape danger?

This **stonefish** escapes danger.

It looks like a rock.

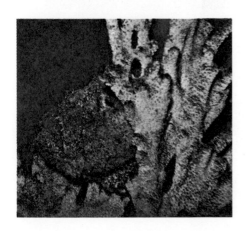

A **chipmunk** can run away to escape.
It can also hide in its hole.

How does this rabbit escape danger?

Animals get food in many ways.

Many animals eat other animals.

Birds have **sharp beaks.**

Some birds can catch bugs.

An **anteater** has a long **tongue.**

Its tongue is sticky, too.

So this animal can catch tiny ants.

Many animals do not catch food.

Their food does not run away.

What kind of food is this?

What animals eat this food?

54

Show what you know

Look at each picture.

How is the animal like an animal you know?

1. This animal moves most like

a _____ .

turtle   chipmunk   robin

2. This animal escapes danger most

like a _____ .

stonefish   chipmunk   cat

3. This animal gets food most like

a _____ .

bird   deer   dog

Ocean life

An **ocean** is very wide.

You cannot see land across an ocean.

An ocean is full of **salt** water.

An ocean is full of living things.

What lives in an ocean?

Plants grow in the ocean.

One ocean plant is called **seaweed**.

What color is seaweed?

Animals live in the ocean, too.

Some ocean animals are very small.

Some ocean animals are very large.

Some **whales** are 30 meters (100 feet) long!

Many animals live in the ocean.

This is a **seashore**.

This is where the ocean and land meet.

What can you find at a seashore?

Some animals live in **shells**.

You can find shells in the sand.

Look for other things.

Activity

Make an ocean picture

Make a class picture.

Paint a big ocean and seashore.

Draw a plant or an animal.

Color it. Cut it out.

Paste it on the class picture.

Show what you know

Which sentences are right?

1. A whale is very small.

2. Seaweed is an ocean plant.

3. Some animals have shells.

4. Ocean and land meet at the seashore.

9

Tracks

Animals make tracks.

Tracks show which way the animal went.

A deer makes tracks in the snow.

Which way is the deer going?

A mouse makes little tracks.

Find its tracks in the snow.

Find its **tail** track.

Big animals make big tracks.

A horse is big.

Find its tracks in the snow.

Are they big?

A **fox** makes tracks in the snow.

The fox came near the horse's tracks.

Find the fox's tracks.

Are they as big as the horse's tracks?

Many things make tracks.

We can find tracks in snow.

We can find them in wet sand.

We can find them on a **sidewalk**.

We can even find them in our homes.

What made these tracks?

Activity

Make some tracks

See how tracks are made.

Get some small things.

Press them into clay.

Take them out.

Look at the clay.

What do you see?

Show the clay to a friend.

Guess what made the track.

Find some tracks.

Find some shadows.

How are they **different**?

Show what you know

What made these tracks?

10

Senses

We have five **senses**.

They help us find out about things.

We see things with our **eyes**.

We see animals.

We see colors.

We see boys and girls.

What can you see here?

You **taste** things with your **tongue**.

You **smell** things with your nose.

What can you taste and smell?

Have a party, and see!

We hear things with our **ears**.

Play a game.

What do you hear?

We **feel** things with our **skin**.

A kitten feels **soft**.

Ice feels cold.

How does water feel?

Activity

Play a game

Put some things in a bag.

Ask a friend to touch one thing.

Ask how it feels.

Ask what it is.

Show what you know

Make a chart about your five senses.

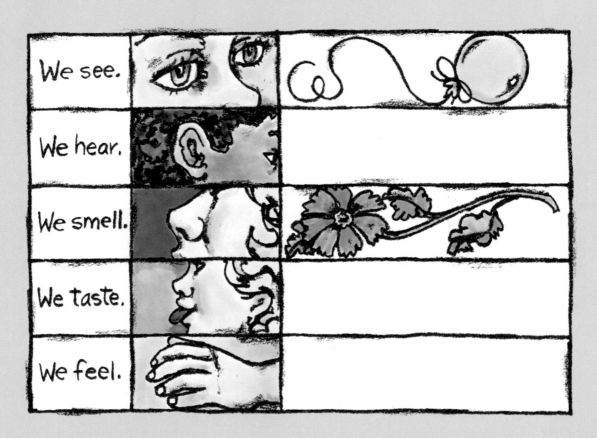

We see.		
We hear.		
We smell.		
We taste.		
We feel.		

11

Feeling good

What do we need to stay **healthy**?

Can you think of some things?

Everyone needs food to stay healthy.

Milk and fruit are good for us.

Fish, meat, and bread are good, too.

Here are some good foods.

Name some good foods that you eat.

We need to **exercise** every day.

Swimming is fun.

It is one way to exercise.

Can you think of some other ways?

We get tired from work.

We get tired from play, too.

Rest helps us **feel** well.

Rest helps us stay healthy.

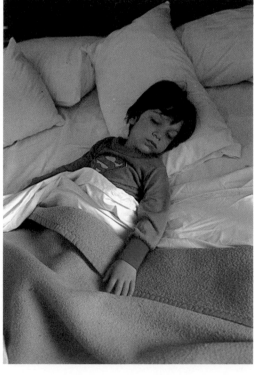

Staying **clean** is good for us.

We wash our face and hands.

We brush our teeth.

This helps us stay healthy.

The doctor helps us stay healthy, too.
Visit the doctor for a **checkup**.

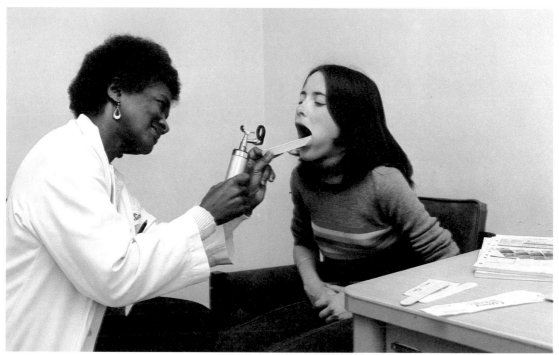

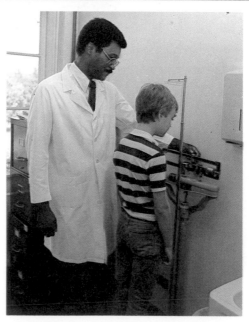

Show what you know

How do you stay healthy?

Draw a picture.

Tell what you do.

12

Safety for you

Do you walk to school?

Do you ride a bus to school?

Do you walk, ride, and play **safely**?

Here are some things to think about.

Streets are for cars, trucks, and buses.

Sidewalks are for people.

Walk on the sidewalk to be safe.

You want to cross the street.

This boy shows you how.

He walks to the corner.

He crosses only at **crossings**.

Be sure the light for you is green.

Be sure the sign says WALK.

Look both ways for cars.

Do you ride the bus safely?

Sit in your seat, and talk **softly**.

The playground is a place for fun.

But sometimes we are not **careful**.

And we get hurt.

Think of some ways to play safely.

Show what you know

Here are some things to think about:

1. You want to cross the street.

   What should you do?

2. You are riding the bus.

   How should you ride?

3. You are playing on the swing.

   How should you swing?

Earth and space science

13. Temperature
14. Water in air
15. Day and night
16. Seasons

13

Temperature

What is the **temperature**?

Is it hot?

Is it cold?

How can you tell?

Look at a **thermometer**.

A thermometer tells if it is hot or cold.

A thermometer **measures** temperature.

Red **liquid** can move up or down.

The temperature gets colder.

The liquid moves down.

The temperature gets warmer.

The liquid moves up.

What is the temperature?

Here is another way to tell.

Look out your window.

What do you see?

Is it hot or cold?

People wear light clothes in hot **weather**.

Some people swim.

Some people lie in the sun.

What do you do when it is hot?

People wear warm clothes in cold weather.

Some people skate.

Some people play in the snow.

What do you do when it is cold?

Some things make the temperature change.

A fireplace can make things warm.

A fan can make things cool.

What makes your home warm or cool?

Show what you know

Make a calendar.

Show the temperature for each day.

Draw pictures to show the weather.

14

Water in air

Water is all around us.

Clouds are made of water.

The water comes from the **air.**

How does water get into the air?

After rain, find a puddle.

How big is the puddle?

Wait for the sun.

Look at the puddle.

How big is it now?

How long will it stay?

What do you see?

What is happening to the puddle?

Make a chart.

Write down what you see.

Time	Size	Deep	Wide
9:00 a.m.			
11:00 a.m.			
1:00 p.m.			
3:00 p.m.			

Activity

Where does water go?

Put some water on a towel.

See where the water goes.

Wait.

Look again.

Is the towel still **wet**?

Where did the water go?

Look at the hot water.

What is happening?

Where is the water going?

Heat helps water get into the air.

Hot air can hold lots of water.

Take three wet mittens.

Put one in the sun.

Shake one mitten.

Leave one mitten alone.

Which **dries** first?

Which dries last?

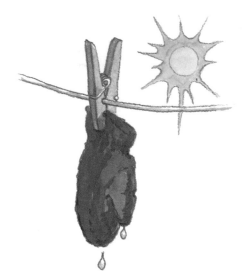

Show what you know

Which sentences are right?

1. Water is in the air all the time.

2. Heat makes things dry slowly.

3. The sun makes a puddle grow bigger.

4. Water from wet things goes into the air.

15

Day and night

In **daytime**, it is light out.

The sun gives us light.

It is easy to see things.

We can see flowers and water.

Sometimes we can see the sun.

What can you see in the daytime?

It is **dark** at night.

The sun is gone.

It is not easy to see at night.

Sometimes we can see the **moon**.

Sometimes we can see **stars**.

What can you see at night?

One **whole** day has day and night.

Daytime begins when the sun **rises**.

What do people do in the daytime?

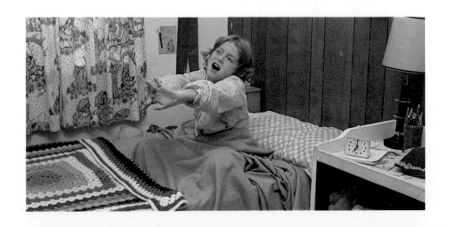

In the evening, the sun sets.

It is now night.

What do people do at nighttime?

After nighttime, daytime comes again.

 **Activity**

Make a chart

What do you do in the daytime?

What do you do at nighttime?

Make a list like this one.

Write a check (✔) for daytime.

Or write a check (✔) for nighttime.

Add other things to your list.

Things I Do	Daytime	Night
1. I eat lunch.		
2. I play with my friends.		
3. I take a bath.		

We see and hear things in the day.

We see and hear things at night.

What things tell us it is day?

What things tell us it is night?

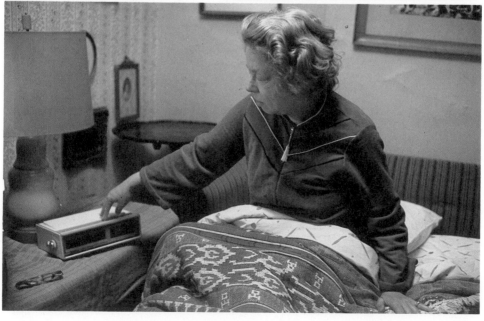

Show what you know

These are pictures from one day.

What is the first thing you do?

Then what will you do?

What is the last thing you do?

121

16

Seasons

There are four **seasons** in a year.

Can you name them?

Winter

In many places, **winter** is very cold.

The days are short.

The nights are long.

Winter may bring snow and ice.

Winter may bring a lot of rain.

What is winter like where you live?

Spring

The days get warmer in **spring**.

In many places, buds grow on trees.

Birds make nests in spring.

What do you do in spring?

Summer

In many places, **summer** is very hot.

The days are long.

The nights are short.

Some people pick flowers.

Some people go swimming.

What is summer like where you live?

Fall

Fall comes after summer.

In some places, leaves change color.

Days get colder.

Animals get ready for winter.

What do you do in fall?

Look at each picture.

Match each season to a picture.

How did you know?

Show what you know

Make a seasons chart.

Season	People	Plants	Animals
Winter			
Spring			
Summer			
Fall			

3

Physical science

17. Mirrors
18. Shadows
19. Who is right?
20. Magnets
21. A closer look

17

Mirrors

What is a **mirror**?

A mirror is **shiny**.

You can see yourself in a mirror.

Sometimes a **window** is a mirror.

It is **dark** outside the window.

There is light in front of the window.

You can see yourself in the window.

Water can be a mirror.

A puddle is a little mirror.

A lake is a big mirror.

When is water a good mirror?

Hold a book.

Look in a mirror.

What do you see?

Does the book look the same?

A mirror can make things look **backward**.

Use a mirror.
Make this look backward.

HO

A mirror can make things look **whole**.

A mirror can make the bread look whole.

Activity

Working with mirrors

Look at these pictures in a mirror.

How does each picture look?

Show what you know

Which sentences are right?

1. A mirror is shiny.

2. A window is always a mirror.

3. Sometimes water is a mirror.

4. A mirror makes things look backward.

18

Shadows

Here are some **shadows**.

What things are making the shadows?

Can the shadows help you guess?

How are shadows made?

You need **bright** light.

You need something to **block** the light.

People and animals can make shadows.

Houses can make shadows.

Windows do not make shadows.

They do not block the light.

Look at these shadows.

What is blocking the light?

Activity

Does a shadow move?

Make your own shadow.

Have a friend draw around it.

Move away.

Look at your shadow drawing.

Where is your shadow now?

Why did it move?

Sometimes you see shadows.

Sometimes you do not.

Why do shadows go away?

We can use shadows to tell time.

Where is the shadow in the **morning**?

Where is the shadow at **noon**?

Where is the shadow in the **afternoon**?

Show what you know

1. Be a shadow maker.

    Make some shadows.

    What is blocking the light?

    Have your friends guess.

2. Make a little shadow grow.

    Make a shadow with your hand.

    Make the shadow bigger.

19

Who is right?

Is it **heavy** or **light**?

Sam says a brick is heavy.

Father says a brick is light.

Who is right?

Which boy on the seesaw is heavy?

Which boy is light?

Is a leaf heavy or light?

Sam says it is light.

To an ant, a leaf is heavy.

Who is right?

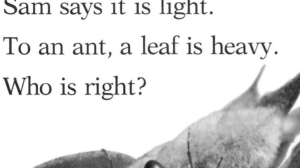

Is a **grasshopper** big or little?

A grasshopper is big to a **ladybug**.

Sam says no.

A grasshopper is little to Sam.

Are two chickens too many?

One family says yes.

One family says no.

Two chickens are just right.

 Activity

Is it big or little?

Make a chart.

Name something in school.

Write it on the chart.

Is it little or big?

Who says it is little?

Who says it is big?

Add more things to the chart.

Something in School	Who says it is little?	Who says it is big?
desk	6th graders	a goldfish
book		
pencil		an ant

The girl looks cold.

The **polar bear** does not look cold.

It does not feel cold.

Why?

Show what you know

Read each word.

See what matches the word.

LIGHT

BIG

HOT

COLD

HEAVY

LITTLE

155

20

Magnets

What can we learn about **magnets**?

Some of these things **stick** to a magnet.

Some do not.

What will a magnet pick up?

A magnet will pick up a pin.

A magnet will pick up a **nail**.

Will a magnet pick up paper?

Try it, and see.

Activity

What can magnets do?

Put two magnets near each other.

What happens?

Magnets can **push away** from each other.

Magnets can **pull towards** each other.

Can one magnet pick up another?

Try it, and see.

Magnets have different names.

Their shape tells their name.

Find a **horseshoe** magnet.

Find a bar magnet.

Find a ring magnet.

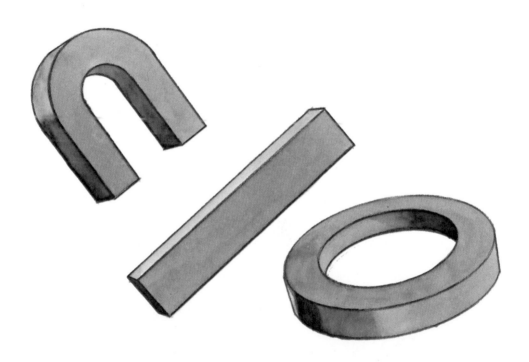

People use magnet **power**.

Look for magnets.

Magnets help people at work.

Magnets help people at school.

Magnets help people at home.

Do magnets help you?

Show what you know

What can a magnet pick up?

Find things in your classroom.

Make a chart.

Objects		Yes	No
1. paper clip		✓	
2. key			
3. tacks			
4. chalk			
5.			

21

A closer look

Find a **leaf**.

Look at it **closely**.

What can you see?

Look at the same leaf.
Use a **magnifying glass**.
What can you see now?

Look at some sand.

Each **grain** of sand was part of a rock.

What color are the grains of sand?

Look at other little things.

Use a magnifying glass.

Little things look bigger.

Activity

Take a closer look

Look closely at some **sugar**.

Draw what you see.

Look at the sugar again.

Use a magnifying glass.

Draw what you see.

Does the magnifying glass help you?

Look at other things.

Look at a quarter.

Look at your hand.

Use a magnifying glass.

See things you could not see before.

Find a **newspaper**.

Find a page with a picture.

Look at the picture.

Use a magnifying glass.

What did you find out?

Show what you know

What is it?

Match the word with the picture.

Leaf

Sand

Coin

Newspaper Picture

Discover more

You can learn more about your world.

Ask your mom and dad to help.

Look at the animal.

Take good care of the animal.

1. Be sure it has food.

2. Be sure it is not too hot or too cold.

3. Be sure it has a clean home.

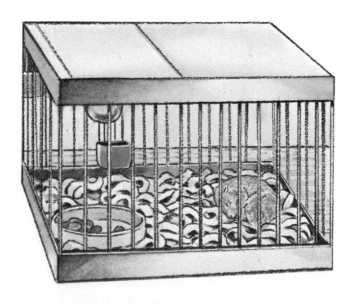

Grow plants.

You need: seeds   water

           soil   sunshine

                 pot

Find leaves, shells, or rocks.

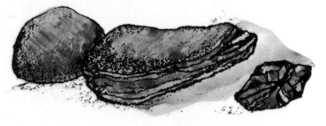

Write down what you see and find.

Vocabulary list

acorns	p. 20	clouds	p. 107
afternoon	p. 146	crossings	p. 91
air	p. 107	dandelions	p. 3
anteater	p. 54	danger	p. 51
aquarium	p. 46	dark	pp. 116, 134
backward	p. 136	daytime	p. 115
beaks	p. 53	different	p. 70
bean	p. 23	dries	p. 112
block	p. 142	earlier	p. 38
branch	p. 34	ears	p. 76
bright	p. 142	escape	p. 51
buds	p. 33	exercise	p. 83
calendar	p. 105	eyes	p. 74
careful	p. 94	fall	p. 127
chart	p. 25	feel	pp. 77, 84
checkup	p. 86	field	p. 3
chipmunk	p. 52	fins	p. 42
clean	p. 85	flower	p. 4
closely	p. 165	fox	p. 67

174

grain	p. 167	mirror	p. 133
grasshopper	p. 151	moon	p. 116
goldfish	p. 41	morning	p. 146
gray	p. 35	nail	p. 158
grow	p. 21	newspaper	p. 170
half	p. 29	noon	p. 146
healthy	p. 81	oak	p. 19
heat	p. 111	ocean	p. 57
heavy	p. 149	pine needles	p. 28
horseshoe	p. 160	polar bear	p. 154
inside	p. 22	power	p. 161
later	p. 34	pull towards	p. 159
leaf	pp. 4, 165	push away	p. 159
light	p. 149	pussy willow	p. 35
lilac	p. 35	rest	p. 84
liquid	p. 100	rises	p. 117
magnets	p. 157	root	p. 12
magnifying glass	p. 166	safely	p. 89
measures	p. 100	safety	p. 89

salt	p. 57	stem	p. 12
scales	p. 43	stick	p. 157
seashore	p. 61	stonefish	p. 51
seasons	p. 123	sugar	p. 168
seaweed	p. 58	summer	p. 126
seeds	p. 6	swim	p. 42
senses	p. 73	tail	p. 66
shadows	p. 141	taste	p. 75
sharp	p. 53	temperature	p. 99
shells	p. 61	thermometer	p. 100
shiny	p. 133	tongue	pp. 54, 75
sidewalk	p. 68	travel	p. 50
skin	p. 77	vegetables	p. 15
smell	p. 75	weather	p. 102
snakes	p. 50	wet	p. 110
soft	p. 77	whales	p. 60
softly	p. 93	whole	pp. 117, 137
spring	p. 125	window	p. 134
stars	p. 116	winter	p. 124

Picture credits

1: Kent & Donna Dannen;   2: Yeager & Kay/Photo Researchers;   4: Grant Heilman Photography;   5t: Hans Pfletschinger/Peter Arnold, Inc.,   5b: S.J. Krasemann/Peter Arnold, Inc.;   6: Jane Burton/Bruce Coleman, Inc.;   7: Frank Siteman/Stock, Boston;   8: Robert Capece/McGraw-Hill;   9tl: S.H. Gottscho/American Museum of Natural History,   9tr: Grant Heilman Photography,   9bl: J.E. Thompson/American Museum of Natural History,   9bc: Eric Crichton/Bruce Coleman, Inc.,   9br: Richard Weiss/Peter Arnold, Inc.;

10: Peter Vadnai/McGraw-Hill;   13tl: A. Singer/Phototake,   13br: Grant Heilman Photography;   14t: Robert Capece/McGraw-Hill,   14bl: Wally McNamee/Woodfin Camp,   14br: Michael & Barbara Reed/Animals, Animals;   15: William Hubbell/Woodfin Camp;   17t: Burpee Seed,   17c: W.H. Hodge/Peter Arnold, Inc.,   17b: John Colwell/Grant Heilman Photography;

18: Grant Heilman Photography;   20L: David Overcash/Bruce Coleman, Inc.,   20r: W.E. Ruth/Bruce Coleman, Inc.;   21L: Grant Heilman Photography,   21r: Runk/Schoenberger/Grant Heilman Photography;   22t: USDA,   22b: Grant Heilman Photography;   24: Barry L. Runk/Grant Heilman Photography;   25: Robert Capece/McGraw-Hill;   26: Clyde H. Smith/Peter Arnold, Inc.,   28: Anita Sabarese;   29: Peter Arnold, Inc.;   30: Robert Capece/McGraw-Hill;   31tl: Earth Scenes,   31tr: Lisa Limer,   31cl: Lisa Limer,   31cr: Anita Sabarese,   31bl: Grant Heilman Photography,   31br: Alan Pitcairn/Grant Heilman Photography;

32: R.V. Boger/Photo Researchers;   34: William Harlow;   35: Grant Heilman Photography;   36: Grant Heilman Photography;   37: Peter Vadnai/McGraw-Hill;   38: Randy Matusow/McGraw-Hill; H. Sund/Image Bank;

40: Robert Capece/McGraw-Hill;   42-44: Peter Vadnai/McGraw-Hill;   45: Yoav Levy;   46: Peter Vadnai/McGraw-Hill;   47: Robert Capece/McGraw-Hill;

48: Elisabeth Weiland/Photo Researchers;   50t: Russ Kinne/Photo Researchers,   50b: Leonard Lee Rue III/Photo Researchers;   51t: Grant Heilman Photography,   51b: American Museum of Natural History;   52t: Leonard Lee Rue III,   52b: Charlie Ott/Photo Researchers;   53tl: Hal Harrison/Grant Heilman Photography,   53bl: Hal Harrison/Grant Heilman Photography,   53r: G. Ziesler/Peter Arnold, Inc.;   54t: Joseph Van Wormer/Bruce Coleman,   54b: Leonard Lee Rue III/Monkmeyer;   55t: Leonard Lee Rue III/Monkmeyer,   55c: Bruce Coleman, Inc.,   55b: Evelyn Appel;

56: Luis Villota/Image Bank;   58: Bob Evans/Peter Arnold, Inc.;   59t: Jeff Rotman/Peter Arnold, Inc.,   55b: Tom McHugh/Photo Researchers;   60t: Richard Ellis/Photo Researchers,   60b: Tom McHugh/Photo Researchers;   61: Russ Kinne/Photo Researchers;   63: Fred Bavendam/Peter Arnold, Inc.;

64: Leonard Lee Rue III/Photo Researchers;   66tl&br: Roger Maserang,   66tr: Tom McHugh/Photo Researchers,   66bl: Evelyne Appel,   68tl: Alan Mercer/Stock, Boston,   68tr: Randy Matusow/McGraw-Hill,   68b: Morton Beebe/Image Bank;   69: Robert Capece/McGraw-Hill;   70: Patrick W. Grace/Photo Researchers;

70: New York Zoological Society;   74: Ginger Chih/Peter Arnold, Inc.;   75&76: Robert Capece/McGraw-Hill;   77: Bill Gillette/Stock, Boston;   78: Robert Capece/McGraw-Hill;

80: Hugh Rogers/Monkmeyer;   82: Runk/Schoenberger/Grant Heilman Photography;   83: Frank Siteman/Stock, Boston;   84L: H. Gritscher/Peter Arnold, Inc.,   84r: Abigail Heyman/Archive Pictures;   85t: Robert Capece/McGraw-Hill,   85b: John Lei/Stock, Boston;   86t: Elizabeth Crews/Stock, Boston,   86b: Peter Vadnai/McGraw-Hill;

88: Robert Capece/McGraw-Hill;   92&93: Robert Capece/McGraw-Hill;   94: S. Sweezy/Stock, Boston;   96&97: Frank Siteman/Stock, Boston;

98: Photo Researchers;   100: Peter Vadnai/McGraw-Hill;   101: Werner H. Muller/Peter Arnold, Inc.;   102: Cary Wolinsky/Stock, Boston;   103: John Colwell/Grant Heilman Photography;   105: Robert Capece/McGraw-Hill;

106: David Hamilton/Image Bank;   108&110: Peter Vadnai/McGraw-Hill;   111: Randy Matusow/McGraw-Hill;   113: Mark Mittelman/Taurus Photos;

114: Doug Wilson/Black Star;   116: Phototake;   117t: Bohdan Hrynswych/Stock, Boston,   117b: Robert Capece/McGraw-Hill;   120tl: Dan McCoy/Black Star,   120tr: Mark Godfrey/Archive Pictures,   120b: Robert Capece/McGraw-Hill;